Glynn Isaac and the Search for Human Origins in Africa

AF864951

Glynn Isaac and the Search for Human Origins in Africa

Rhys Isaac

Inaugural Glynn Isaac Memorial Lecture

21 April 1993

UCT Press (Pty) Ltd
University of Cape Town
P/B Rondebosch
7700 South Africa

Published in Association with
the Department of Archaeology,
University of Cape Town

Copyright: Rhys Isaac
University of Cape Town
First published: 1995
ISBN: 0-7992-1587-2

All rights reserved. No part of this publication
may be reproduced, stored in a retrieval system,
or transmitted in any form or by any means,
electronic, mechanical, photocopying, recording
or otherwise, without prior permission of the publisher.

Typeset by Tina Barsby
Printed and bound by Creda Press, Cape Town

Contents

List of Illustrations

Glynn in his element—showing what a stone blade can do

Preface

Glynn Isaac was born in Cape Town in 1937, and died in Tokyo in 1985. He was too young to die. Not yet fifty, he had a wife and two teenage daughters he adored, a fabulous new job at one of the world's finest universities, and a new house in walking distance of his department. He was at the apex of his professional career, an established professor, brimming with enthusiasm for the scatters of rock and bleached bone that encode the origins of humanity.

His death was perceived immediately as a shocking loss to the profession. Hundreds gathered at memorial services in Cambridge, UK and Cambridge, Mass., where normally composed colleagues wept unashamed from the podium. His students and friends convened to analyse his intellectual legacy at conferences, each holding a strand of the large multi-disciplinary venture that characterised his research projects. Barbara, his wife, gathered her strength and embarked on the major project of bringing his Koobi Fora research together in a single volume.

In the tea room of the Department of Archaeology at the University of Cape Town, where it all started so many years ago, it came to a small band of his friends that, for all these fine tributes, a lasting and continuous reminder of the man also would be fitting. We had in mind something that would recur again and again as a vehicle to remind us and so many others, not only that we loved him, but that his peculiarly generous intellectuality still inspires us today. We also wanted something that would help inspire African students to join in the study of human origins—a goal which Glynn embraced throughout his career.

Hence, the Glynn Isaac Memorial Lecture, the first of which was held at the University of Cape Town on 21 April, 1993.

It was a dramatic step, but hardly overdrawn, for Glynn's career abounded in bold steps. He came early to archaeology, a barefoot schoolboy, scouring the sites at the Cape of Good Hope. From there he moved north, less into the heart of darkness than into the rising light of early hominid studies of Olduvai Gorge, Olorgesailie, and Koobi Fora. There his flair and intellectual breadth bloomed, as he lassoed other disciplines into his ambit, bringing a variety of workers from Africa and abroad, to address broad-ranging issues of site formation, taphonomy, and prehistoric human social organisation.

The intellectual agenda that he laid down then is still being actively pursued, like the meaning of scatters and patches in an archaeological landscape, which received extensive attention at the most recent Paleoanthropology Meetings held in Oakland in March, 1995. In a seminar here at the Department of Archaeology in Cape Town in 1979, Glynn epitomised our earliest ancestor with a slide of a black "Eve", thus preceding his Berkeley colleagues' rendition of a Mitochondrial Eve by a good five years.

It was fitting that Glynn's earliest and possibly best friend, protagonist, and intellectual sparring partner, his twin brother Rhys Isaac, should give the first Glynn Isaac Memorial Lecture, and that the subject this time was Glynn himself. In the future, we will invite distinguished scholars in the field of Palaeoanthropology (broadly defined, to include the archaeological and biological evidence for human origins) to give a popular lecture on current developments in the discipline. In this way, we hope to stimulate interest in the study of human origins, both here in the city which gave Glynn life, and farther afield.

We honour Glynn's work, his intellect, his breadth of vision, and his memory in the first of what we hope will be a long series of Memorial Lectures.

John Parkington
Carmel Schrire
Andrew Sillen

Cape Town, 1995

And Death shall have no dominion . . .

Here I stand back at UCT, in awe of Glynn's work—of its range, reach and implications for science, for humanity, and for those who are from and of Africa. Yet I am proud to speak, in this memorial lecture, for Glynn's many friends and admirers here, and for his fond, proud and still sorrowing family on four continents.

From my own situation as twin—sharing so much from our origins here, sharing a reverence for the Africa that gave us birth, and for the human past which humankind must seek to understand—I shall try to present a view of Glynn's work and his legacy as I see it.

In October 1978 I received a letter from Glynn from the wilds of east Lake Turkana. He wrote to me about a moving moment in his lifelong quest to decipher the record concerning humankind's origins—concerning "the evolution of human behaviour", as he usually termed it. This was a moment brought about by the generosity Glynn inspired in those who shared his quest:

> Let me go on to tell you something of our safari to Laetoli . . . For some years Mary Leakey has been seeking to get me to come down to her work at Laetoli 30 miles south west of Olduvai, where they have been finding 3½ million year old hominid fossils and in the last years some layers with animal footprints. Always I have left it to the end of our season & have been caught up—unable to go . . . This year Mary told me that they had found a trail of hominid footprints, but that they would have to cover them over at the end of Sept. because of the rain . . . so I decided to play hookey from here for 10 days, drive down, pick Barbara up in Nairobi, & go to Tanzania . . . .
>
> Laetoli turns out to be a beautiful little valley complex at the foot of the Ngorongoro Volcanic highlands—on the edge of the Serengeti . . . fine—but seeing the footprints was an utterly extraordinary, moving experience . . . There one stood looking across 3½ million years at a specific moment in proto-human life.
>
> So far they have traced a trail which with small breaks goes back 23 metres. There are two parallel trails, both unmistakably human . . . one

> large—walking with parallel tread & a stride of 50-55 cm and another smaller, walking with splayed tread.
> The trails are so close that they must either have been walking with locked arms, or one half behind the other . . . The smaller, presumably female, trail-maker stopped at a certain point and turned half sideways to look at something . . . ! An amazing record to have been preserved, saved through time, then found and studied . . . .

Let me share also passages from two other letters of Glynn's—his last from Africa to his family in Australia. The first is dated from "under the thorn trees at Kampi ya Kivuli," 16 August 1985—some six weeks before the terrible fever carried him off. ("Kampi ya Kivuli"—meaning Camp of Rainclouds—was the archaeologists' "home base" beside a dry stream bed out in the arid wilderness, miles back from Lake Turkana and the big HQ camp at Koobi Fora.) Glynn's letter opened:

> Here I sit with my back against an *Acacia tortilis* trunk, its reclining branches making a tracery against the blue evening sky. Golden light flows up the sand river bed in front of me [,] and on the opposite bank a huge *Delonix* tree rises out of the greenery and is a shower of golden blossom. The acacias are more discreetly in flower—but enough to attract delicate little sunbirds . . . Off to my left a mile or so away is the Karari scarp with our beloved sites. Off to the west where the sun is setting is the lake & the Koobi Fora base camp—for us but a way station to this, one of our spiritual homes.

He went on to declare how "it is always soul restoring to return to Africa & to this place".

Of course Glynn's engagement with that landscape was not all, or even primarily, romantic sunset-on-the-African-veld-type writing. Far from it. Five years earlier, in a wonderful rigorous team-work scientific essay on one of those "beloved sites" (appropriately entitled "FxJj 50") he had limned that same terrain very differently, the UCT-trained geologist in him reading it persuasively in terms of slow processes of a distant past reversed to make a new set of slow processes in the present. I quote from the article:

> Flying north-eastward from Koobi Fora, one leaves the shores of Lake Turkana and heads across low-lying country towards a sinuous line of hills. These hills form part of the rim of an ancient sedimentary basin into which river systems have been discharging loads of gravel, sand, silt and clay for several millions of years. In many places these layered deposits preserve a fossil record which includes both palaeontological and archaeological evidence relating to very ancient proto-human ways of life . . . .
>
> Today deposition has largely ceased, and the layered deposits that were laid down between 1 and 3 million years ago are being eroded. Under the . . . arid conditions which now prevail, areas of bush and scrub are interspersed among badlands where erosion gullies form intricate dendritic patterns . . . . (Isaac [1980], 1989, 228)

In an almost-last letter—a follow-up one, two weeks after the sunset one already quoted, a letter written on an aeroplane as Africa receded tragically for the last time—Glynn resumed for us the Kampi-ya-Kivuli theme:

> We had a good bout of field work [he wrote] . . . &, as always, I found that this spell of living in a vast empty, open landscape, of walking among thorn trees & over outcrops, of watching sunrise & sunset make the *Acacia tortilis* umbrellas where we camp into the most delicate lattices against the glow of the sky . . . all this really gives me a lift and restores the soul—in addition to which we got valuable research done . . . .

And he recorded that he was able to help with a new field-school venture in which African and American students, side by side, got "instruction & mild apprenticeship in the geological, palaeontological, and archaeological components of human-evolution field studies." What is more, Glynn had been able to strengthen one of the most valued partnerships of a working lifetime of collective endeavours. Richard Leakey had flown Glynn back from Koobi Fora in his light aircraft, and put him up for his stopover in Nairobi. "This gave me a chance," Glynn wrote, "to see him at leisure & his family . . . He and I have the deep friendship of shared adventures and of having been beginners together."

I have given you a few precious moments captured from Glynn's writings—a moment of pleasure in contemplating partnership in adventure in wild African fastnesses, a moment of satisfaction in inter-racial teamwork in a human-evolution field school in that same open veld landscape; a moment in an aeroplane analysing scientifically the terrain of the Rift Valley below; the evocation of evening light in the thorn trees as a way of sharing known-and-loved experience with his distant family; and—supremely—Glynn's moment of communion with a 3½-million-year-old little episode in the emergent human past. These moments all evoke stories, as well as suggesting *the* story of Glynn's questing across the far wide African veld, and across the far wide, largely unmapped veld of early almost-human time.

Being human we tell stories—indeed, in some sense, we are, if not the stories we tell, then the stories that we carry and that we continually enact. As a historian I have come to be fascinated by all the making of stories, the leaving behind of stories, and the taking up of stories that is history. I am fascinated by the stories that come down to us from past people, and also by the stories that we historians—indeed, we humanity—make of the stories that come down to us. I have here collected some stories of Glynn, and I have attended closely to the stories that Glynn collected for us. He told powerful stories in his vital writings—stories of an intensive search, and stories of the paths we, humanity, have travelled to become human. For this occasion, therefore, I have tried to assemble a story of Glynn's journeying forth from this place of his birth and upbringing and education—his journeying forth to make his great contributions to African studies and to a humane science of human origins.

Stories of Glynn

Perhaps this is not the time and place for a lot of stories recalling Glynn the person. In any case I have a difficulty here with regard to the beginnings. For our early, formative years we shared everything—including identity. We wore clothes interchangeably from a common wardrobe, we shared all possessions, activities and friends. (Even when we went to UCT and embarked on separate courses, and each acquired distinctive sports jackets, we still—at night before we went to sleep in our shared two-bunk room out the back at La Rochelle on Camp Ground Road, Newlands—pooled our now-separate experiences of the day.) Around the UCT campus we each untroubledly greeted persons we did not recognize; *we* knew who these persons *thought* they were greeting.

Our habit of pooling experience upon getting together once again produced at least one memorable episode. Glynn was at an archaeological excavation, I had just arrived to join him—so, naturally I just dumped my gear and dropped into his trench so we could catch up with each others' doings. After a while, Jim Sackett, a now very distinguished archaeologist who was excavating in the next trench, jumped up and shouted, "Stop it, you're driving me mad!" Our voices being indistinguishable, what he seemed to hear was one person give a longish anecdotal recital and then answer himself from a slightly shifted position with an equally long reply. And so, the fact is that having grown up so intertwined, I don't have a stock of stories of early Glynn distinct from myself—only stories, like the above, of our togetherness.

Before I leave the theme of sharing that formed our earliest years, and that entered so deeply into both Glynn's way of working and into the story of human origins that he dug—as we shall see—from the earth, I must say a word about the loving wisdom of our parents that sustained us as the inseparable twins we were. Much should be said to recall here what they gave us—our mother, Frances Leighton Isaac, the Bolus Herbarium flowering-plant botanist, and our father, William Edwyn Isaac, the marine ecologist, and much-loved Harry Bolus Professor of Botany at UCT. I can only allow myself to tell one thing and that must stand for the whole. Our parents suppressed and kept as a close-guarded secret the information as to which of us was born first. That seems a small thing—but it required great determination. The world seeks incessantly to divide and to impose hierarchy. People constantly confronted our parents and us with the questions: "Which one is the elder? Which was born first?" It was a great reinforcement of our sharing—a powerful avoidance of the division that these people unthinkingly sought to introduce—that we could always answer truthfully: "We don't know. Our parents didn't separate us that way". As I have said—loving wisdom—and a great foundation for some of what ran deepest in both Glynn's contribution and the way he came to make it.

Despite our early substantially merged identity, Glynn's path and mine had begun to diverge, our lives, that is, had begun to separate, long before we left home to go for further study overseas at separate universities, long before we began, for the first time in our twenty-one years of life, to live apart. That divergence had, in fact, begun on a much earlier journey overseas. Aged nine, turning ten years old, we both responded strongly to our first exposure to Old-World antiquities when we went with our mother and father, on his first return to his native Wales after the war, in 1947. Glynn was so taken with the treasures of the British Museum that he resolved to be an Egyptologist—an archaeologist therefore. I was so taken with castles and cathedrals that I wanted to be a historian. (I still in my heart of hearts tend to think of medieval history as "real history", and to identify myself as a historian of the early modern period, seeing my studies in the Age of the Enlightenment as marking somehow, most of all, the "real" end of the Middle Ages!)

Glynn acquired a manual of Egyptian antiquities, by James Henry Breasted. I can still see him poring endlessly over this fat book, and I am glad to say, Barbara still has it among her most treasured relics of Glynn and his life's work. In time—I do not remember distinctly when—Glynn shifted his interests to South African prehistory and Burckitt and Van Riet Lowe became his bibles. His zeal and the keenness of his eye gave rise to an incident vividly recalled in family folklore. Our father, Edwyn, likes to tell how all doubt in his mind as to whether Glynn really meant it when he said he would be an archaeologist, was dispelled by an episode one time as the family was out on a drive. Edwyn recalls how Glynn, aged 15, called out, as the car sped down the road, "Stop, I see an Acheulian hand ax!" We stopped, we reversed a way, and there it was! If he could recognize its type in a heap of roadside stones flashing by at 45 mph, Edwyn was sure Glynn *must* be destined for discoveries in archaeology.

Carmel Schrire, fellow Cape Town archaeologist, has a recollection from that time. Carmel recalled being with her father on a South African Archaeological Society outing to a midden and artefact site by the shore:

> being about nine, I was more entranced by shells. My father frowned on this frivolity. "Look at that boy," he hissed, "he 'knows' implements, and look how he concentrates." The boy in question was small, with light crinkly hair, dirty knees and a Fair-Isle jumper. From the way in which his smiling father, the Professor of Botany, and [the senior archaeologist] John Goodwin treated him, he seemed to 'know' things . . . I imagine scowling. But within a week I mounted a set of silcrete microliths on a cardboard boxtop, redeeming myself a little in my father's eyes. (Schrire, 1985)

A little later, Carmel and myself formed Glynn's first excavation crew, when we went one very wet, late-winter weekend to excavate the bleak hand-ax littered surface of Cape Hangklip. There we dug out what I believed were the only 30 cubic metres of hard Cape soil anywhere without a single stone in it—while Glynn discoursed to us increasingly forlornly on "the value of negative evidence"!

Glynn's Journey Forth

How should the story of Glynn's life work be told? There are so many ways that needs to be done. No doubt a specialist in each of the many sciences Glynn drew on and contributed to would see it and tell it differently. But looking from my historian's perspective, I see in Glynn's life's work that there was indeed a comprehensive concern that I can take up to direct my telling. Both our working lives—Glynn's and mine, that is—have been given to searching out, finding and learning to read records of a distant past—and to telling and teaching as we went, what we found and how we read the record. Distance, however, is relative. I have searched at two centuries ago; I stand in awe of one who peered back into the records of two million years. My narratives of Glynn's journeying will be of his determined quest to find readable records of emergent humanity.

Coming back now to the place of our beginning, remembering our time as students at UCT—Glynn in the Science Faculty, majoring in Zoology and Geology (with all the Archaeology then available added), I in the Arts Faculty, majoring in History—I can see that what must above all be told is the extraordinary way in which Glynn combined science and arts for all of his so creative career onwards from UCT. Sciences gave the framework and the method; the search for humanity was always the objective. Glynn applied scientific observational and experimental approaches across an astonishing range—and he gathered teams of students and colleagues to join him in that work and to pursue the possibilities to the limits. They entered the engineering sciences of hydraulic flows, the ecological sciences of carnivore and scavenger behaviours, and of habitat-specific availabilities of plant foods; they entered experimentally the artisanal domain of basalt stone-tool making, and of butchery with the products. (Indeed, they were so keen to test the efficacy of the tools that they put out word to all zoos. When an elephant died at Denver Zoo, Glynn's team rushed there to butcher the creature—successfully, as he related to the Royal Society in his address for the Darwin Centenary.) In all this teamwork—behind Glynn's encouragement of all this experimental scientific observation—there was a humanist impulse, an impulse to know proto-humans as far as possible in the way only humans can know humans, that is by their actions and their utterances, and the traces, the human record, those actions leave.

Enquiry into human meanings and intentions is not possible on the basis of the palaeolithic record. Enquiry into behaviours is. Glynn moved determinedly into behaviour studies—especially adaptive behaviours with selective advantage in the evolutionary process. He insisted that the archaeology of human evolution must play a central role in the fast developing sciences of the evolution of mankind.

We stood at the start of this hour with Glynn looking at the Laetoli footprint trail, so we have shared a sense of the strength of his craving to find communication with the human in the distant past. But that Glynn would not do self-indulgently as a sentimental projection of the present onto that past. In his considered contributions to the discourse of his field, every conclusion was critically assessed and from preference cast as a hypothesis to be falsified or as one of a set of alternatives for testing. The UCT-trained historian in me responds deeply to Glynn's determination throughout his career to recover "readable" records of very early emergent human behaviour; the UCT-schooled scholar in me is in awe of the rigour of his equal determination to search out ways to test experimentally the reading of such a problematic record.

Glynn's deep desire was to get critical knowledge of what it meant to experiment scientifically, for ways to reconstruct the behaviour of those earliest almost-human beings and for ways to test scientifically those reconstructions. Always it was a quest for humane knowledge of *who* we humans are, out of knowing the way we have come.

I shall follow in sequence the main steps of Glynn's journeying on from UCT—the steps of his quest for records of early humanity and for the sciences by which they might be read.

From UCT Glynn went in 1959 to Cambridge—but before he could go, his character and the quality of the preparation UCT had given him, was to be proved, when his mentor, John Goodwin, found himself dying of lung cancer and unable to conduct his classes. Glynn, just 21 years old and newly graduated with a plain BA, was called on by Professor Monica Wilson, whose protegé he also was, to give the lectures, and to conduct the classes. He rose to the occasion and managed wonderfully. Arriving at Cambridge with such proven strength, he rapidly won the respect of his archaeology teachers there, making a lifelong friend of the great Charles McBurney, and winning himself an Elsie Ballot scholarship to see him through. The same proven character and determination as well as intellect enabled Glynn to go on from Cambridge in 1961 to his first appointment back in Africa, and there to triumph over difficulties that had defeated others.

In Kenya he was able to seize his first chance to recover and interpret a very early and very problematic record of early human life, at and around the former Lake Olorgesailie in the Rift Valley, some 50 kilometers southwest of Nairobi. (I still have the most wonderful letters from Glynn dating from this exciting place at this heady time.) At Olorgesailie Glynn also took the next and most important and rewarding step of his career after leaving Cape Town. Barbara Miller came to join and to marry him, to live and work with him in a little cluster of grass huts on the side of a great dry river bed across from the old volcano of Mt Olorgesailie. It was a lifelong partnership that sustained Glynn literally to the end, and that is still continuing, as Barbara, even now—overcoming all the difficulties besetting such an operation—brings out the culminating Koobi Fora monograph of Glynn's research team's collective work. Together in that wild African setting they learned to live and work together and to be good neighbours to the people all around. It was to a hilltop near Olorgesailie that Barbara, with their daughters, Ceri and Gwyneira, went to scatter Glynn's ashes at a simple rock marker placed there by a group including members of his former excavating team.

Olorgesailie was Glynn's place of apprenticeship. There, with Louis Leakey and Mary Leakey as inspiring, challenging—though not closely supervising—masters, Glynn was curator of an open-air archaeological museum site, and the determined trainer of wage-labourers who must be taught to become an effective team for excavating other sites. If he could rise to those challenges, he could learn to make himself the archaeologist he wished to be, and to enter upon helping remake archaeology as the key to the evolution of humanity he wished it to be. Technical mastery is an essential part of apprenticeship—and I sense, without being able really to judge, that Glynn entered upon that, and attained it superbly. I cannot therefore follow closely the steps here—but I see that Glynn gained the authority both as a field worker and as an assembler/analyst of complex data, to command attention for his work.

Palaeolithic archaeology up to that time had been very largely a matter of stone-tool morphology and classification. Glynn entered that with a thoroughness that enabled him to come out with a serious critique of the adequacy and viability of such taxonomy as the primary concern of researchers in this field. He directed attention *from* putative genealogies of tool types *to* the sites themselves as "documents"—the available record of early humanity. Initiating the close interdisciplinary and cross-disciplinary partnerships that so characterized his career, he worked to show how sites could be "read" as records, on the one hand of the preferred habitats of early humans, and on the other of revealing events and episodes in their long-ago lives.

In directing his efforts thus toward reconstructing past life, Glynn was one of the first to realise the need for an experimental scientific knowledge of the steps by which the underlying events that the archaeologist seeks to know become part of the record available at a site—the ways that events such as the discarding of a stone implement or the dismemberment of an animal carcass leave an enduring trace. Put another way—he began experiments in what is now the great science of taphonomy, as he sought to know what might interfere with, or rearrange, discarded stone artefacts or bones (Isaac, [1967], 1989, 31-57). These experiments had an elegant simplicity that deserves to be celebrated here in a review of Glynn's work that emphasizes the ways he combined science and arts, at once testing hypotheses and searching for contacts with humanity in the past.

Glynn had been fortunate, but he also used his situation well. His apprenticeship at Olorgesailie had been in what Glynn himself declared to be the seemingly "monotonous", unglamorous, Middle-Pleistocene. He had worked on the long period from about a million to about a hundred thousand years ago, when change was scarcely detectable in the record, and only the most meticulous work could earn attention and respect. At the same time he had been close to the most momentous developments in his field—close, that is, to the work of Louis and Mary Leakey that had consolidated the claim (laid decades previously by Raymond Dart and others in South Africa) for Africa to be recognized as *the* place of human origins. Between 1959 and 1961, the work of the Leakeys at Olduvai Gorge in Tanzania had transformed the field by picking up crucial new hominid fossils—both of the genus *Australopithecus* already known from the Transvaal, and of the then earliest species of our genus, *Homo habilis*. Even more momentous for Glynn, the Leakeys had found those fossil remains (dated at about 1.8 million years ago) associated for the first time with crude but unmistakable stone tools.

In 1966, after some four years of field-work apprenticeship and a year as a graduate student back at Cambridge, Glynn went to the University of California at Berkeley. In the Anthropology Department there he associated closely not only with the redoubtable Desmond Clark—with whom Glynn had already determinedly gone to work in Zambia as an undergraduate during one UCT summer vacation—but, also, among others, with the great student of ape behaviour, Sherwood Washburn, and with the geologist responsible for clarifying the Olduvai sedimentary sequences, Richard Hay. The setting was right to broaden the enquiry—to go, that is, from the search for "missing links" on the fossil-hunting side, and from the exquisite distinguishing, classifying, and computation of stone tool assemblages on the archaeological side, onwards towards the systematic study, as Glynn used to put it, of "the evolution of human behaviour" (Isaac, [1972] (1989), 18).

By 1970 Glynn (with his PhD gained for his thesis on Olorgesailie) was ready for what proved to be the great research project of his life. At this time he renewed an early partnership with Richard Leakey in order to search out the dramatic traces of an early proto-human story that was eroding out along the eastern side of Lake Rudolf (soon to be renamed Lake Turkana). Glynn, with his base well-established at Berkeley, had now a place—a landscape of the past re-emergent into the present—truly to stir not only his own imagination but that of the students who would come to him, and the fund-granting institutions that would enable him to develop a wide-ranging exploratory and experimental programme, and to keep a team in the field to open up that archaeology of human origins.

The direction that enquiry in early hominid archaeology must take, had by this time become clear to Glynn through the revelations coming out of the close behavioural observation of the great apes—most notably out of the work of Jane Goodall on the chimpanzees of Gombe Stream. An overview of the features of chimp social life, ecology, and means of subsistence invited a corresponding overview of what we ordinarily take for granted about humans. Glynn's programme for an archaeology of human evolution that was both scientific and humanistic arose from the contrast, then, between the essential features of the chimpanzee way of life and of the human way. The contrast can only be stated here in bold outline, yet it must be outlined in plotting Glynn's work, because it establishes what he understood to be the beginning of the story of the evolution of human behaviour, and what he perceived as the end result observable among humans in our own world.

The most important differences between humans and the great apes may be, as Glynn liked to point out, not so much anatomical as behavioural. Humans pride themselves on their reason—but underlying that is the intensified, complicated sociality of language and culture. The precondition of that complex sociality is a fundamental shift that has taken place in collective relationship to the means of subsistence. The great apes of Africa—the chimps and gorillas—have lively social relationships that satisfy real needs, but, once weaned, each chimp finds and eats its own food in a way that is not the dominant pattern in any human society. As the chimp band moves through its terrain, each member helps itself to the edibles encountered, consuming as it goes. Even when a number of chimps combine (as they do when the opportunity arises) to capture a small creature, they do not share their quarry among themselves; at most the one in possession grudgingly gives out portions to placate others who come to beg—"tolerated scrounging", Glynn called this. With humans it is otherwise; eating-the-meal is always (even in Bali) overwhelmingly an occasion at the end of complex exchanges and divisions of labour (Isaac, [1978], 1989, 289-311).

Among living peoples who rely heavily on hunting and gathering, there is typically an age and sex division of labour. Greatly simplifying, it may be said: the mature males go hunting or fishing, while the young and the females go gathering vegetable foods. The complex of behaviours that sets humans apart—as Glynn came to point out insistently—depends on two complementary arrangements—"the home base" and "food sharing". Division of labour requires—if some are going to go out to procure one kind of food and others another—that they have an agreed place to which both will return for the exchanges that motivate their postponement of consumption and their effort to transport back what has been procured.

The task of the archaeologist of human origins, as Glynn had come to see it very clearly, is to pick up the traces of the steps, as it were, from the ancestral ape way of life to the hunter-gatherer way that has been directly observed in our own world. From those traces the archaeologist can try to document the circumstances—the places and the changing environment, for instance—and the stages of behavioural adaptation. It is a daunting task—requiring all the tenacity and spirit of scientific enquiry, the readiness to formulate hypotheses, test them, reformulate or discard them, and begin again, that Glynn had developed. The task is especially daunting because the enquirer picks up the story somewhere way past the middle, and, without knowing what came before, must be very uncertain as to just what is being picked up.

The traces of the steps along the way towards humanness are not just stone tools, yet the archaeologist is largely dependent on stone tools as the tell-tale markers of places where hominids once were active. But hominids had come a long way—on two legs already for two or three million years—before they began, at about two-and-a-half million years ago, to make and leave those tell-tale markers.

Apes—it has been shown in the case of chimpanzees—use tools; but *dependence* on tools for survival itself is a distinctive feature of the human way. Glynn often used to quote the insight of Richard Borshay Lee, the famous worker among the Kalahari !Kung San, that *the* indispensable "tool" for the human way of life is "the basket" (or other equivalent forms of carrying containers) (Isaac, [1978], 1989, 306). But, in terms of hunter-gatherer division-of-labour exchanges, the knife, or the tools by which hunting weapons are made, and by which the carcasses to be fed on are divided up, is equally indispensable for the maintenance of the system. Those cutting devices of stone, enduring virtually forever under the right conditions, are thus the widely dispersed markers of places of ancient human activities. Human? or proto-human? That was the problem facing Glynn and his fellow archaeologist searchers after the evolution of human behaviour—seekers for even the faintest traces of those steps by which primal bipedal apes must have begun to evolve our human home-base-food-sharing social subsistence pattern. Basket equivalents vanish leaving no record. Only the stones endure—and, unfortunately, even notoriously, there is nothing inherently social about the knife or the chopper, except the idea and the craftskill to make it.

The challenge—so Glynn had come to realise, as he and his team tried to decipher the million-and-a-half-year-old record eroded and carefully excavated out of the scarps and ravines east of Lake Turkana—was to try to envisage not-yet-human beings with a collective way of life somewhere between ape and human—a way of life that has no observable equivalent anywhere in the present-day world. The search would have to be for records of that way of life with sufficient resolution to give clear indications of habitat, ecology, and patterns of subsistence. An intense experimentation was also needed in order to be able to read those records with a maximum of assurance and a minimum of the self-delusion into which the intense interrogation of the past so readily lapses.

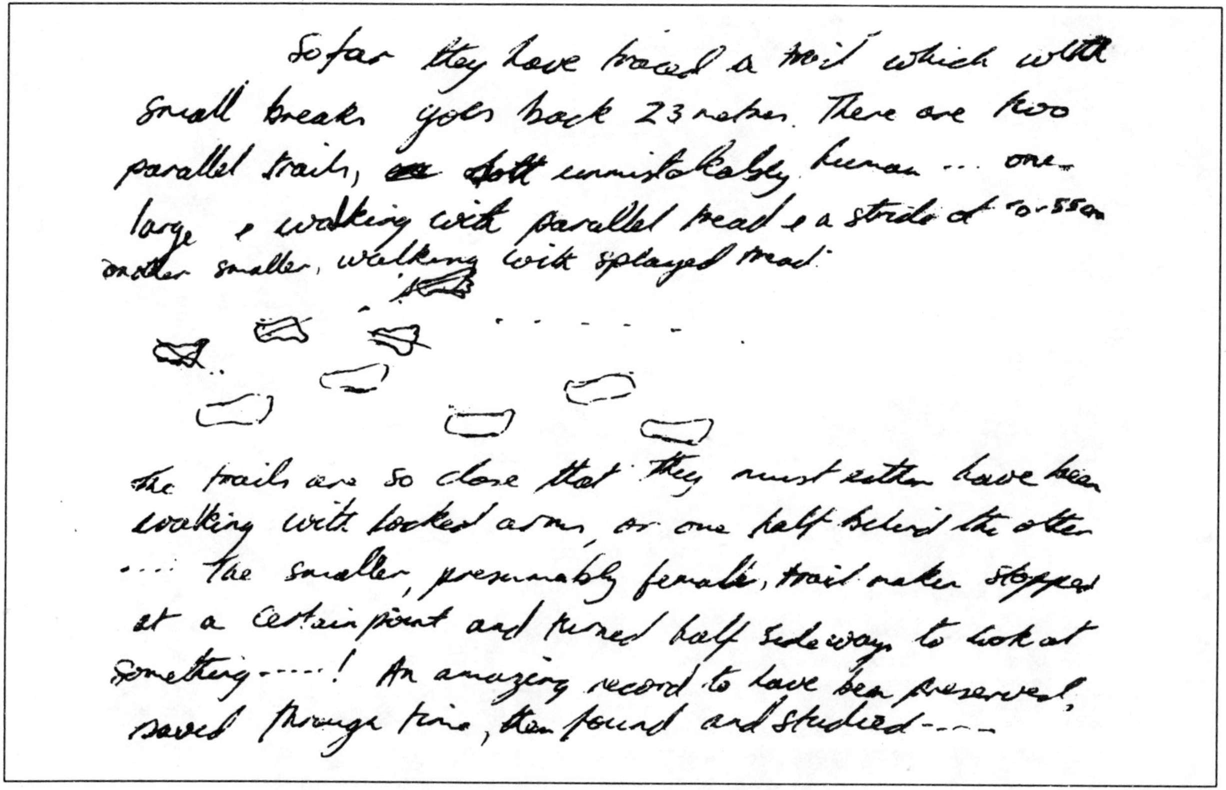

So far they have traced a trail which with small breaks goes back 23 metres. There are two parallel trails, unmistakably human ... one large & walking with parallel tread & a stride of 50-55cm another smaller, walking with splayed tread.

The trails are so close that they must either have been walking with locked arms, or one half behind the other ... The smaller, presumably female, trail maker stopped at a certain point and turned half sideways to look at something.....! An amazing record to have been preserved, saved through time, then found and studied----

Glynn's record of the fossilised moment at Laetoli

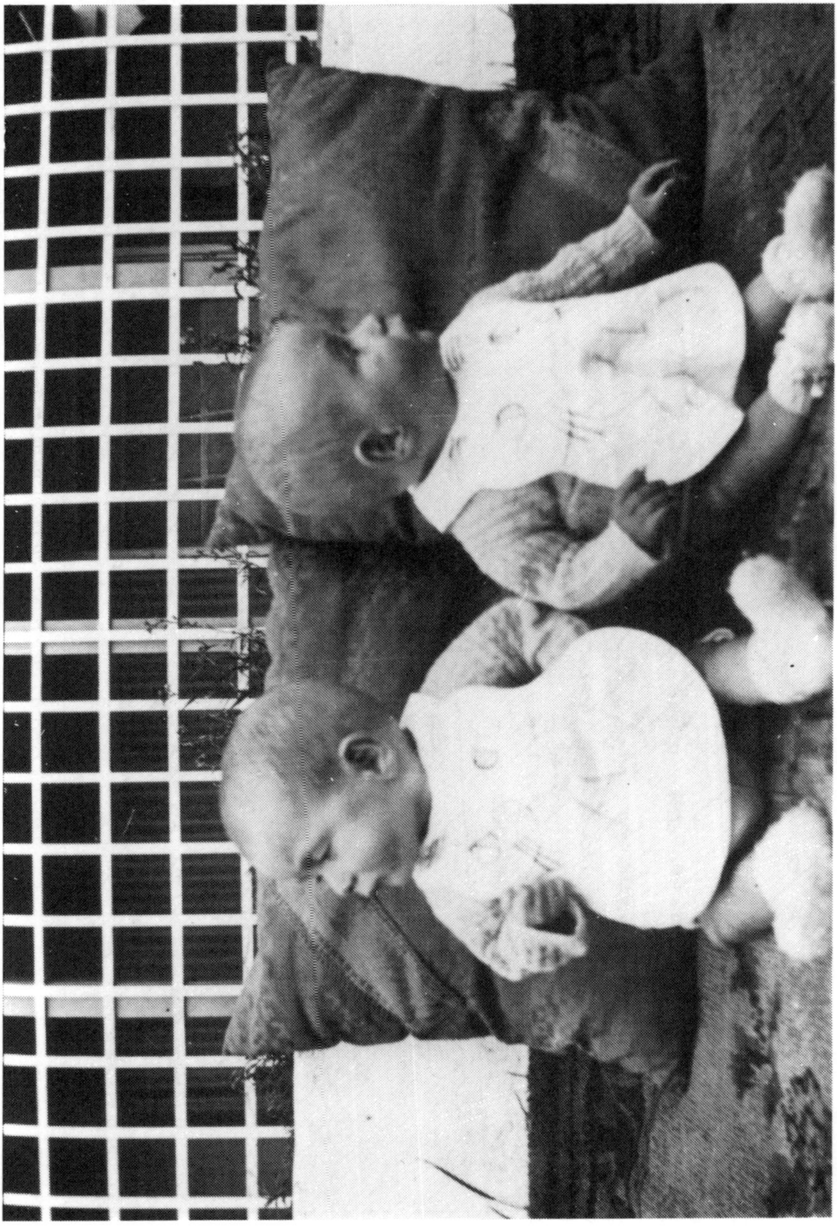

Mirror imaged identity

Edwyn and Frances and the Botany Building—together at UCT in the 1950s

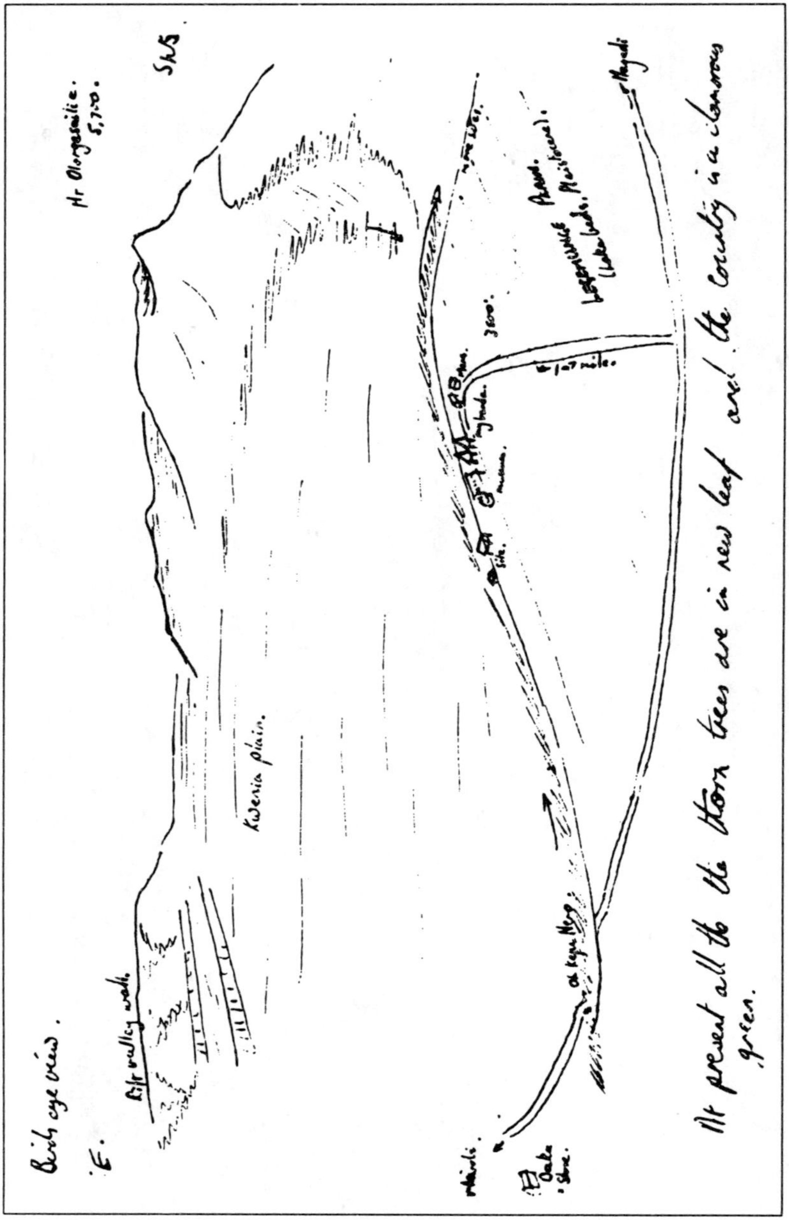
Birds eye view.
E.
Rift valley wall.
Mt Olorgesailie.
5,700.
Kwenia plain.
Magadi
At present all the thorn trees are in new leaf and the country is a glorious green.

A place of origins: Glynn's sketches of the Olorgesailie basin and the home base there

Partnership: Glynn and Barbara recording the Olorgesailie Museum Site

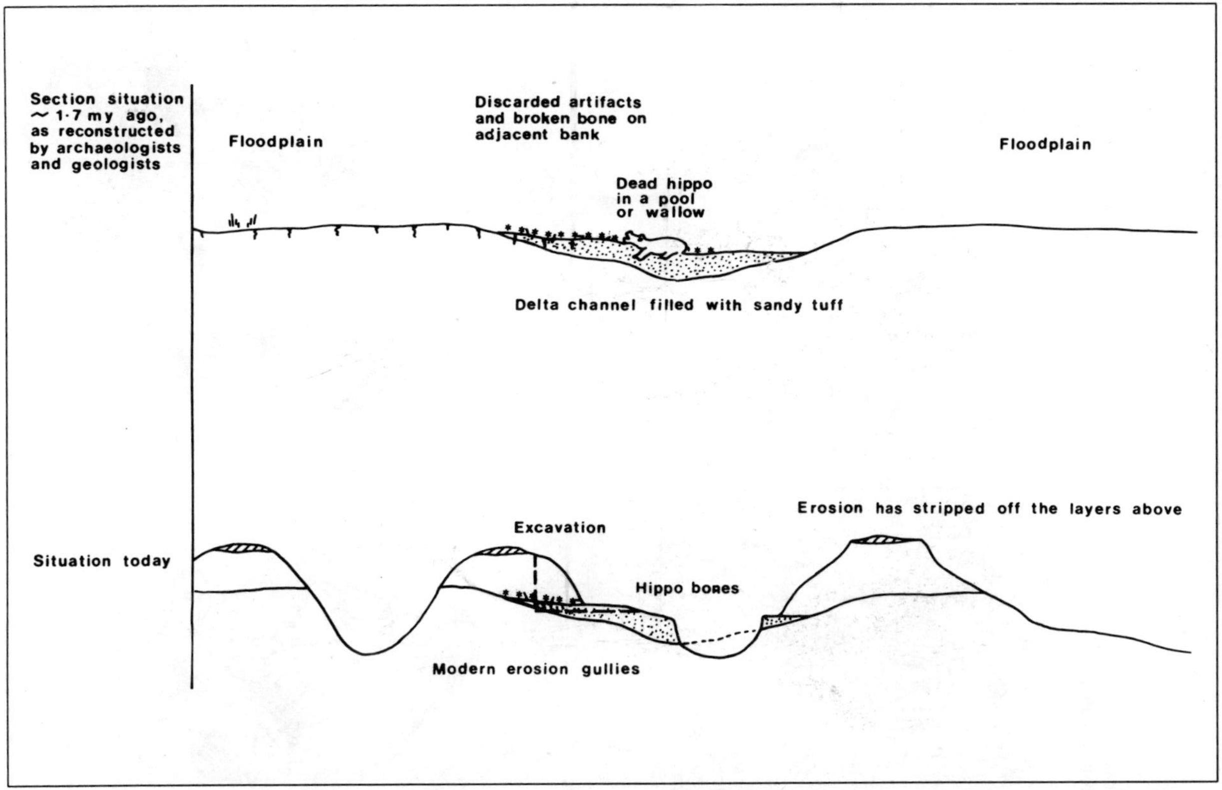

HAS: Hippo and Artefact Site—as formed (reconstruction) and as excavated

Ape's eye view of our contrasting ways of subsistence (Courtesy of Jeanne Sept)

"Beginners together"—Glynn and Richard Leakey in the field

Origin Myth—the KBS figleaf

Excavating FxJj50

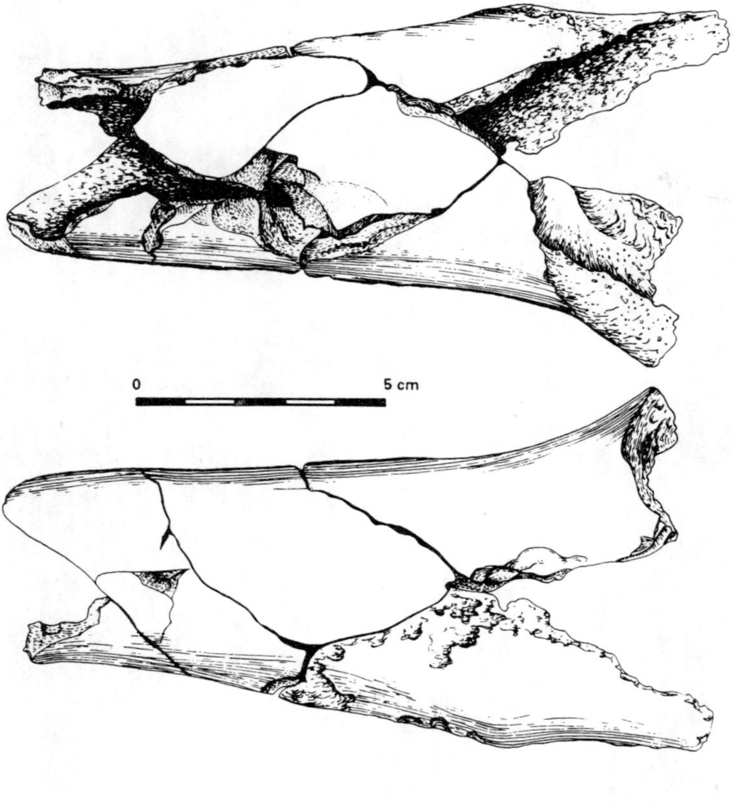

Events in replay: conjoined marrowbone from FxJj50

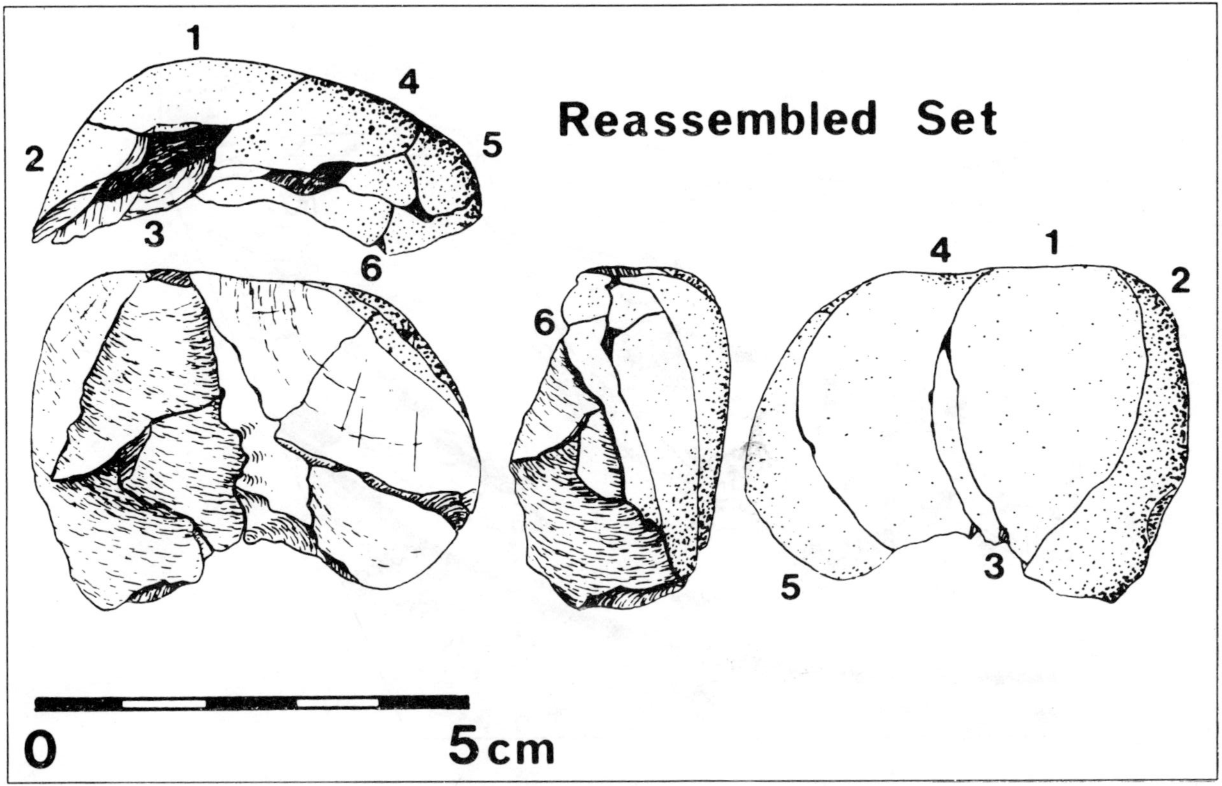

Events in replay: conjoined stone core from FxJj50

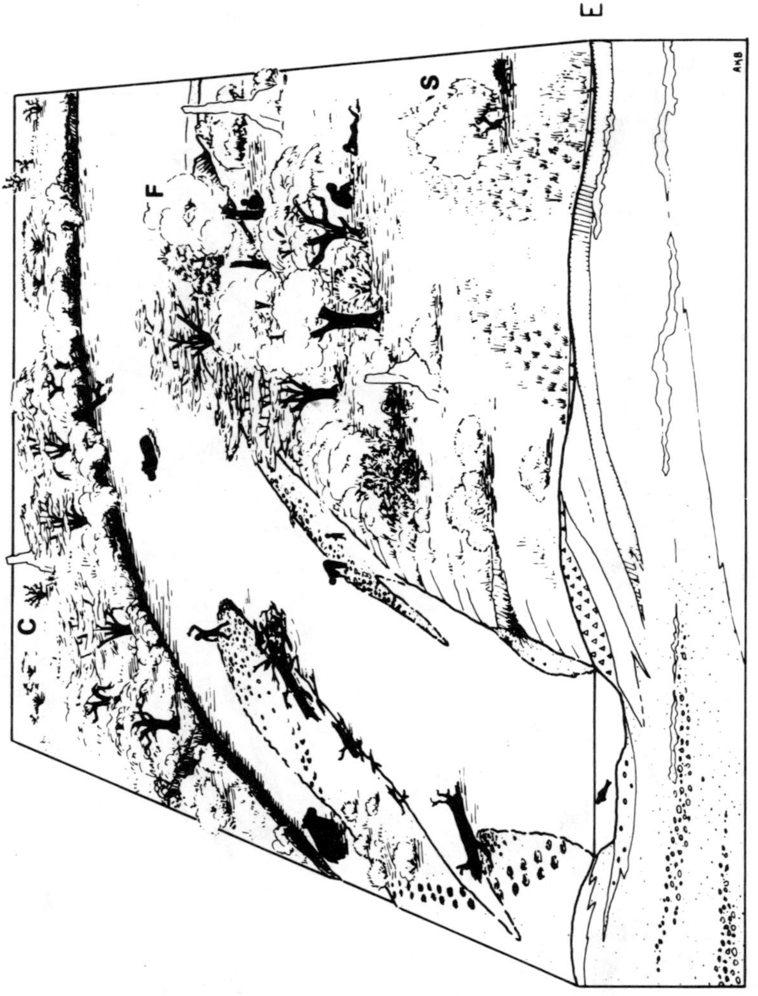

The riverbend site as it might have been: FxJj50 reconstructed

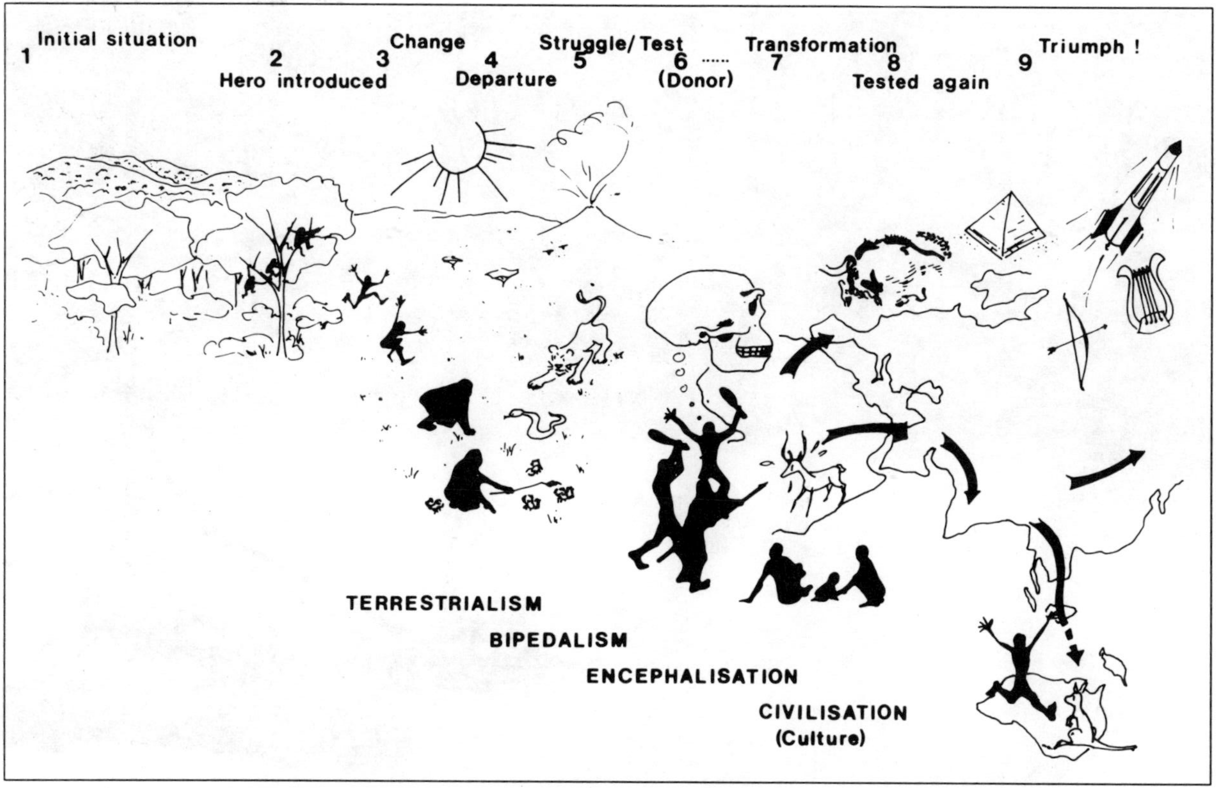

Fun with hypotheses: whimsical sketch of origins (Courtesy of Jeanne Sept)

Toward Olorgesailie—Glynn's all-time favourite picture

Glynn—we shall see—was to have two distinct attempts at envisaging the not-yet-human way of life whose traces he and his team searched and puzzled over so carefully. Through this he also came to have enduring insights into the nature of the challenge to all who enquire curiously into the distant past, seeking to understand extreme "otherness".

The First Round

Two sites became the prime records upon which Glynn based his first attempt to identify and interpret a stage on the way between ape behaviour and human behaviour. At the HAS site (now known as FxJj 50)—the Hippo and Artefact Site as it was first called—the bones of a single large animal (the hippo that gives the name) were found eroded out, and among them were many very primitive but unmistakable stone artefacts. When the ancient topography was reconstructed, it became clear that the hippo had died in a pool or wallow in a sandy-bottomed seasonal channel. It appeared that the carcass had been butchered there by hominids who brought in stones (much larger than any that occurred naturally anywhere within miles of this place). It was thought that the makers of these primitive choppers and stoneflake knives probably had not had the means to kill so large an animal, and that they were, in this recorded episode, scavengers rather than hunters. Nevertheless, the evidence seemed clear. "At least 1.6 million years ago . . . the behaviour of some hominids . . . differed from the behaviour of modern great apes in that . . . [they] not only made cutting tools but also ate meat from the carcasses of large animals" (Isaac, [1978], 1989, 301).

A second excavated site seemed to provide a record from which even more significant readings could be made. It was called KBS at first, the initials standing for Kay Behrensmeyer Site, since it had been affectionately named for the geologist/palaeontologist who found it. At this place there appeared a different configuration from that found at HAS, the Hippo and Artefact Site. Here were intermixed concentrations of stone artefacts and bone fragments from various animals—at least six very different kinds, ranging from porcupine to hippo and giraffe. It was argued that this number and variety of creatures could not have been killed or died all on the one spot, and so it seemed that bones, bearing meat and marrow, had been carried there by the toolmakers. As the team reconstructed this million-and-a-half-year-old topography, the hominids seemed to them to have camped and feasted on the sandy bed of a seasonal watercourse under the shade of trees. Surviving prints of leaves flattened in the mud show that some of those were wild fig trees (Isaac [1978] (1989), 303). (This intrusion, as it were, of sacred story into archaeological record, is noted with amused irony in Glynn's published reports, and it always brought a special smile of delight to his face when he talked about it.) The whole configuration was taken by Glynn to be "highly suggestive, and . . . indicative of a camp or home base, with recurrent dispersal and reunion of the group at the chosen locality". He declared that, "among living primates, this pattern . . . is distinctive of man", and reasoned that the coincidence of bone food refuse with the artefacts strongly implies that meat was carried back—presumably for sharing" (Isaac, [1976] (1989), 87).

At the two key sites, then, there appeared to be clear indications of a way of life that incorporated some of the most distinctive elements of human, as opposed to ape, behaviour; here, it seemed was a fossil "home base" where "food sharing" was practised within a pattern of division of labour. From the evidence that meat was carried repeatedly to the one place—that spot in the sandy dry river bed under the shade of fig trees—it was inferred that vegetable food was also brought there for exchange, although characteristically it had left no trace. From the extremely rudimentary unstylised character of the stone tools found, Glynn inferred, however, that the formal systems of language and culture, and so of full "humanıty", were not yet developed.

Here then was a foundation hypothesis for the new-fledged interdisciplinary field of the archaeology of the evolution of human behaviour—the scientific field that Glynn was doing so much to establish with this work. As a branch of evolutionary zoology/ anthropology, the new field was oriented to issues of adaptive advantage within a process of natural selection. Glynn postulated that these East-Turkana hominids—amongst the earliest to leave an archaeological record of their behaviour—had adapted and survived by developing—through food sharing—a kind of collective responsibility for each other's welfare. Since present-day hunter-gathering societies show a sexual-basis for the division of labour underlying their food-sharing practices, Glynn also postulated that some incipient pair-bonding (of a kind not found among chimps) may have begun to be incorporated in the proto-human way of life that he saw documented in the records at HAS and KBS (Isaac, [1978] (1989), 310-11). Glynn had indeed been for some time an effective critic of the macho-masculinist "man-the-hunter" theory of human evolution: he had insisted on a properly bi-gendered account of the processes (Isaac, [1972] (1989), 18-19).

The archaeology-centred science of the evolution of human behaviour now had effectively a whole new systematic framework for exploring the biological, natural-selection advantages that might have been given by the intelligence of which humans are so proud—the trait whose advance is presumably indicated in the fossil record by the fairly steady enlargement of the cranial capacity of hominid skulls that has occurred across the time-span since stone-tool making began. Individuals with the intelligence to assimilate and communicate subsistence information from the environment—so Glynn reasoned—would contribute substantially to the viability of their codependent group; individuals with the intelligence to manipulate relationships within such a group would make out better and reproduce more successfully—or so he suggested.

The Second Round

Scientific hypotheses are for the testing. The home-base/food sharing hypothesis was no exception. Questions arose about whether the stone tools and bone fragments were so intimately connected as to point unmistakably to a period of time when the hominids who made the tools had brought meat to the one place and created a mixed litter of discarded stones and bones. The uncertainties arising from a record dating back more than a million-and-a-half years, and possibly confusing together the remains of many events occurring in those almost unimaginable aeons of to-us-undifferentiated time—all these uncertainties inevitably haunted the record that Glynn and his team were recovering, and so they haunted the hypotheses that were founded upon that recovered record. Glynn's responses to the challenges that were coming from his own questioning approach, as well as the polemic arguments of others, were two-fold—the one extensive, the other intensive.

The extensive response—about which I am qualified to say little—was to break determinedly out of the mesmerism that "the site"—the putative past settlement place—has long had over archaeologists; he would take the whole landscape as a field of study. Glynn would view the "sites" as merely "patches" within a much more diffuse "scatter"—and ask questions about the whole record that it all constituted. Or rather—such is the plight of archaeologists—since whole million-and-a-half year ago landscapes have either been washed away or buried, he would look at the whole sinuous traverse of the ancient but now buried landscape, that the line of erosion along a scarp could reveal. This was what Glynn had learned as a geology student at UCT, very ingeniously applied to reading the human record. But I shall not here pursue further this innovative systematisation of approach to the Pleistocene archaeological record (Isaac, [1981] (1989), 258-268).

The intensive response is best illustrated by the team-work excavation of a site called, with studied scientific neutrality, by its grid reference; "FxJj 50". This was truly one of the foremost of Glynn and his associates' "beloved sites". The members of the team—each of them already engaged in their own wider-ranging mapping, excavating, and experimental projects—all agreed also to bring their particular developing expertises to bear on this one site. Since the reading of human characteristics into Pleistocene hominid behaviour was being contested, there was an extra-wary thoroughness built into the programme. The questions asked were the crucial ones: was this—as appeared from the limited quantities of stone and bone—a record of a distinct, short-lived and fairly well-defined set of events? How had the associated stone and bone come to be both concentrated and preserved? "Did the concentrations . . . imply that the toolmaking hominids were transporting meat (and other food) to special places (home bases or living sites) for sharing?" (Bunn *et al* (1980 in Isaac, 1989), 228-257, quotation, 230).

The team was not disappointed in the choice of site. They found and, in finding, created a remarkable record, to be read in many complementary ways by means of different expertises. For the team, Kay Behrensmeyer and Z. Kaufulu were able to reconstruct the ancient setting—once again the site was associated with a river channel; Jack Harris and Glynn analysed the stone-tool assemblages; Henry Bunn found butchery cut-marks on some of the bones; Henry Bunn, Ellen Kroll and Kathy Schick were each able to show in different ways how well-preserved had been the integrity of the site at the time it was overlaid with silt. (So gentle had the covering-over process been that certain hammer-shattered bone fragments and knapped stone cobbles could be re-assembled to produce the whole from which the fragments had originally been broken at the site.) Nick Toth showed that the stone fragments (with accountable variations) matched the array of fragments that his own experiments had generated in the here-and-now by knapping equivalent basalt cobbles. All in all, the team could, with a glow of pride, report a kind of meticulous cumulative scientific arrival at an experience equivalent to Glynn's communion with the Laetoli track makers:

> We have had a privileged sense of gaining glimpses of particular moments in the lives of very early proto-humans who lived in Africa 1.5 million years ago. For instance: finding . . . bone with marks apparently formed when a sharp-edged stone implement was used to dismember an antelope leg, cannot but conjure up very specific images of butchery in progress. Finding the fitting pieces of hammer-shattered bone shafts invites one to envisage early proto-humans in the very act of extracting and eating marrow . . . The discovery of clusters of conjoinable flakes establishes a sense of connection across time. (Bunn *et al*, (1980, in Isaac, 1989), 256)

The just-quoted 1980 account of this most "beloved" of their sites was deliberately descriptive and analytic only of what had been found; it wisely did not enter into the by-then controverted interpretation of such findings in an overall account of hominid behavioural evolution. That form of engagement *had* to follow, however, since Lewis Binford's sharply polemical book, *Bones: Ancient Men and Modern Myth*, appeared in 1981, and attacked palaeoarchaeologists in general, and Glynn (via Mary Leakey) in particular, for what were denounced as their unexamined assumptions. Binford declared that the heroic hunting which had been imputed to early hominids was a fantasy imposed by the present on the past—though this was an idea already powerfully worked out in a fine study of mythic assumptions by Timothy Perper and Carmel Schrire (1977). Binford had also claimed to demonstrate that animal activities and non-human agencies could better account for the bone concentrations interpreted as hominid "living sites" by Mary Leakey, and that, therefore—since hominids had not been *shown* to transport meat in significant quantities—the home-base food-sharing hypothesis could be dismissed as a sentimental anachronism.

True to his own need to subject humanistic insight to scientific testing, Glynn had already become deeply engaged in seeking ways to examine his models. He had, in any case, always advanced his interpretations only as a line of investigation, a source of further observation and testing. So, Glynn's responses were entirely characteristic. He had, back in 1974, deplored the tendency by which palaeoanthropology had "come to resemble the stamping ground of rutting male antelopes"—a reproof that may help to account for the ferocity of the stamping that followed (Isaac, [1976] (1989), 347). (See also Binford, (1989), 272-273) In 1981 he tried to keep the whole contest in an appropriately rational mode when he celebrated the signs that a long-standing palaeontologists' passionate tendency to advocate one or another account of human origins was giving way to "the formulation of alternative-hypotheses, each with its own test implications to guide research." Of this shift, he remarked, revealing himself clearly, that "this not only turns out to be a constructive move, but also it is fun!" (Isaac, [1981] (1989), 315). From then on his presentations of the overall interpretations of hominid-activity sites were always in terms of the assessment of alternative possibilities for explaining the observed facts. (The fun to be had from this eschewing of dogmatism in the face of such old, blurred and uncertain records became clearer in a wonderful presentational collaboration he developed with Jeanne Sept, whereby she rendered the various hypotheses and stories concerning hominid adaptive behaviours as whimsical sketches.)

Glynn came, in this process, to lay increasing emphasis on the *non*-humanness of the million-plus-year-ago toolmaking hominids (268, 116). He conceded that he had at first "made the early hominids seem too human", and he came to advocate that the more neutral terminology of "central place foraging" be adopted in place of "home base and food sharing"—since the only immediate inference from observation was that hominids *were* transporting meat and bone (whether hunted or scavenged) back to a preferred place (Isaac, [1984] (1989), 149). (On this last point he was vigorously fighting back, since he insisted that the inference from archaeological data that meat *was* transported *had* withstood the onslaught—the "beloved site", FxJj 50, had shown that.) A possible millions-of-years-long, very slow transition from ape to human ways had, however, begun to emerge from the discussion. On the basis of Rob Blumenshine's close observations of scavenging in the veld, it was suggested with great cogency now, that meat may have been transported to trees for security from big rival scavengers rather than for the sake of sharing. In the fossilisation process, of course, the trees were long gone and the bone and stones alone remained in the ground. Under that view, forms of sharing exchange between meat getters and plant gatherers may have been a much later incidental flow on from the use of certain convenient trees as shelter (Isaac, [1984] (1989), 176).

In short, Glynn's last great creative contribution was to search the record more earnestly—even as he presented possible interpretations modestly and playfully—for the lineaments of a distinctively *non*-human hominid form of adaptive subsistence. He realised both the necessity and the inherent difficulties of this task. His final substantive publication on the subject was in the form of a tribute to his lately deceased teacher, Charles MacBurney—a piece sadly published only after Glynn's own death. In it he offered profound allegories to counsel all who look into the inevitably mythic past. Without any fanfares of post-modernism, he engagingly presented those kinds of insights into the inescapable problems of construing the text, interpreting the record. He reminded palaeoanthropologists of how Christopher Columbus, arriving on the margins of a New World, could only interpret it in terms of the Old World from which he had come. Even more telling, he likened us all to one who looks into a deep well: "beyond the dimly lighted upper rim is darkness extending away from the watcher . . . but in these depths is the gleam of light on the water. If the well is not too deep, by straining our eyes we perhaps see a figure—a figure set in an unfamiliar context but yet a familiar figure—familiar, because it is our own reflection". Indeed, he goes on, including his own hubris in the general chastening he narrated: "archaeological inquiry into the very remote past has recently experienced comparable moments of truth." To these humanistic insights, he has a scientist's questioning response:

> Can our scientific imaginations rise to the occasion? Can we conceive of patterns of behaviour and adaptation that could lead to the formation of familiar-looking patterns of archaeological evidence, and yet were behaviours unfamiliar to us in that they were structured differently from recent . . . human ones? Can we then figure out predictable differences in the archaeological evidence that will distinguish the various possible behavioural models? (Isaac, [1986] (1989): 378).

When—at our last meeting, our last mountaineering excursion together—Glynn explained to me what was happening to his view of the hominid past, the humanist historian in me regretted the substitution of "central place foraging" for "home base and food sharing". The humanist in me still regrets the shift, but I understand better now the resolute endeavour behind it—the endeavour to imagine a way of being not only other than our own but nowhere observable in the world about us. In any case the original insight did not go away, and I can still rejoice in being the twin, and—in common with this whole assembly here—a Cape Town compatriot of the one who so resolutely (and correctly) identified being and becoming human with acts of sharing. Glynn had come to realise more fully at the end of his career just how long and slow—lasting way past the "invention" of stone tools perhaps—was the way from being ape, and feeding oneself alone, to being human. But the lesson remained: the attainment of full humanity means sharing space in home places and sharing food in versions of the meal—

with all the sacred obligations of restraint and cooperation that this places on us. Truly a lesson of the past for our present times.

Archaeology is profoundly an Arts/Humanities enquiry—especially when it is—as Glynn came increasingly to realize—mythically engaged in a quest for origins.

Archaeology is profoundly a natural-science enquiry—especially when you give yourself, as Glynn did, a UCT-science (zoology and geology) background for it, and commit yourself to the interdisciplinary enhancement of a palaeoanthropology that observes and experiments searchingly to learn to read the deeply obscure millions-of-years-old record of a deeply obscure proto-human way of life.

Archaeology not only seeks a record in the traces that remain, but it creates a record by its carefully noted excavations and experiments.

Archaeology in creating its record, creates places—places where the landscape of the present, touches and intersects, as it were, with the landscape of the past. At these places—their "beloved sites"—by rigorous experimental science, the team could find and share a kind of connectedness, a strenuously disciplined version of the Laetoli moment through which we sense communion with the past.

Glynn's inspirational influence—his deeply mourned charisma in palaeoanthropology—was derived from here, in the shelter of Table Mountain, where he and I were born, at the tip of Africa. From here he went far—not just in physical distance, but in journeys of the spirit, back in time. His high intellect, the calibre of his science, the quality of his humane insights, won him great honours—all of which should be regarded as honours for the place from which he started. He became a distinguished professor first at the University of California at Berkeley, and then at Harvard; he addressed the Royal Swedish Academy of the Sciences, and gave the Kroon

Lecture at the Hague. Glynn's eminence in human evolution studies was honoured in his prominence at the Royal Society's celebration of the centenary of Charles Darwin. Yet the secret of Glynn's charisma lay, not in ambition for such honours, but in the richness and the generosity with which he combined three qualities: a humane engagement with the story of human origins—an inescapably mythic story—(remember his delight in the fig leaves); his equal commitment to scientific experimentation, as both a conduit of imagination and the test of truth; and through it all, his identification with African landscape—settings to be shared by all African peoples.

Africa is indeed a place. Glynn contributed mightily to the great collective work of palaeontologists and archaeologists (and others) by which Africa has become *the* place of human origins. (When Glynn was starting out, that was only the implicit claim of Raymond Dart and Robert Broom—a claim little regarded in the big world of Europe and North America, where they do not expect great truths to come "from the South".) During Glynn's working lifetime, not only was Africa established as *the* place of human origins—but the sense of what those origins must have been was transformed. Raymond Dart had sought the key in hunting and blood lust; Glynn learned, by contrast, from the African apes and those who watched them, and he learned from the hunter-gatherer peoples of Africa and those who know them, that the key lies in *our* human ways of cooperating—combining to make places together (home bases) and sanctifying our collective lives in those places by taking meals together (food sharing).

Glynn should be remembered and the co-operative spirit of his work continued in a search for human and proto-human story, in a rigorous scientific extension and control of that search, and in a reverence for African landscape as the portentous setting for the origin places of human cooperativeness, human sharing. I hope that the University of Cape Town will be supported in its intention to further such purposes by establishing an annual Glynn Isaac Distinguished Lecture here at the place where Glynn was given his start.

References

Binford, L.R. 1981. *Bones, Ancient Men and Modern Myths*. New York: Academic Press.

—. 1989. *Debating Archaeology*. San Diego: Academic Press.

Isaac, G.Ll. 1989. *The Archaeology of Human Origins: Papers by Glynn Isaac*. Edited by Barbara Isaac, Cambridge: Cambridge University Press. (Note: in references to this collection I have shown the original date of publication in square brackets).

Perper, T., and C. Schrire, 1977. The Nimrod Connection: Myth and Science in the Hunting Model. In *The Chemical Senses and Nutrition*. Edited by M.R. Kare, New York: Academic Press.

Schrire, C. 1985. Memoir of Glynn Isaac. Unpublished contribution to the Glynn Isaac memorial service, Peabody Museum, Harvard University.

Selected Annotated List of Publications of Glynn Isaac

Note: Most of the works listed below have been conveniently assembled in one volume by his career-long co-worker Barbara Isaac, with an introduction by John Gowlett: *The Archaeology of Human Origins: Papers by Glynn Isaac*. Edited by Barbara Isaac, Cambridge: Cambridge University Press. 1989. This work also contains a complete listing of the published writings of Glynn Isaac, pp xvii-xxv.

1967 Towards the interpretation of occupation debris: some experiments and observations. *Kroeber Anthropological Society Papers* 37:31-57.

Very important experimental work in what would be the science of taphonomy, undertaken during Glynn's apprenticeship years at Olorgesailie and published when he went to work with J. Desmond Clark at Berkeley.

1971 The diet of early man: aspects of archaeological evidence from Lower and Middle Pleistocene sites in Africa. *World Archaeology* 2:278-99.

The themes of landscape, environment, and subsistance that dominated in Glynn's work are already clearly stated in what was also a balanced, implicit critique of the sexist Man-the-hunter account of human evolution.

Whither archaeology? *Antiquity* 45:123-9.

A prize-winning essay defending new trends in scientific archaeology against old-guard humanist complaints.

1972 Chronology and the tempo of cultural change during the Pleistocene. *Calibration of Hominid Evolution: Recent Advances in Isotopic and other Dating Methods Applicable to the Origin of Man*: 381-430. Edited by W.W. Bishop and J. Miller, assistant editor S. Cole. Edinburgh: Scottish Academic Press.

A bold synthesis of knowledge in its day—but still illuminating—concerning the time scales of the human past.

Early phases of human behaviour: models in Lower Palaeolithic archaeology. *Models in Archaeology*: 167-99. Edited by D.L. Clarke. London: Methuen.

An important methodological essay in an important methodological book.

1976 The activities of early African hominids: a review of archaeological evidence from the time span two and a half to one million years ago. *Human Origins: Louis Leakey and the East African Evidence*: 482-514. Edited by G. Ll. Isaac and E. McCowan. Menlo Park: Benjamin.

A systematic probe into the earliest traces of proto-human ways in a volume produced by Glynn in tribute to his great mentor, and later identified by Glynn as the place where he first advanced the concept of home base and food sharing in combined formulation.

Stages of cultural elaboration in the Pleistocene: possible archaeological indicators for the development of language capabilities. In *Origins and Evolution of Language and Speech*: 275-88. Edited by S.R. Harnad, H.D. Steklis and J. Lancaster. New York: New York Academy of Science.

An exciting attempt to make seemingly unpromising records yield clues as to the timing of the development of the untimely human activity of languaging.

1977 *Olorgesailie: Archaeological Studies of a Middle Pleistocene Lake Basin in Kenya*, assisted by Barbara Isaac. Chicago: Chicago University Press.

The great monograph in which Glynn's apprentice fieldwork was fully published and his determination shown to have archaeology reveal way of life in reconstructed environments.

1978 The food-sharing behavior of protohuman hominids. *Scientific American* 238 (4):90-108.

The food-sharing and home-base hypothesis of human evolution presented systematically out of a comparison of ape and human subsistence patterns and a reading of two of the most important lower Pleistocene sites in Glynn's Lake Turkana work.

1980 With H. Bunn, J.W.K. Harris, Z. Kaufulu, E. Kroll, K. Schick, N. Toth and A.K. Behrensmeyer. FxJj 50: an early Pleistocene site in northern Kenya. *World Archaeology* 12:109-36.

Co-authored by the team, this must surely be one of the most searching as well as dramatic readings of an early human record.

1981 Archaeological tests of alternative models of early hominid behaviour: excavation and experiments: *Philosophical Transactions of the Royal Society of London*, series B, 292, 177-88.

An intellectually honest review of the ways to pursue enquiries into rival hypotheses concerning early hominid behavioural evolution.

1981 Stone Age visiting cards: approaches to the study of early land-use patterns. In *Patterns of the Past: Studies in Honour of David Clarke*: 131-55. Edited by I. Hodder, G. Ll. Isaac and N. Hammond. Cambridge: Cambridge University Press.

A brilliant example of Glynn's application of his flair for testable hypotheses, to the reading of palaeoenvironments.

With J.W.K. Harris and F. Marshall. Small is informative: the application of the study of mini-sites and least-effort criteria in the interpretation of the early Pleistocene archaeological record at Koobi Fora, Kenya, *Las Industrias mas Antiguas*: 101-19. Edited by J.D. Clarke and G. Ll. Isaac. Mexico: X Congresso Union Internacional de Ciencias Prehistoricas y Protohistoricas.

A very important scatter-between-the-patches hypothesis for systematising the total archaeological record given one of its first explicit treatments.

1983 Aspects of human evolution, *Evolution from Molecules to Men*: 509-43. Edited by D.S. Bendall. Cambridge: Cambridge University Press.

An overview of theories of the evolution of human behavior from Darwin to the present, appropriately presented at the Darwin centennial.

Bones in Contention: competing explanations for the juxtaposition of early Pleistocene artifacts and faunal remains. *Animals and Archaeology, Hunters and their Prey*: 3-19. Edited by J. Clutton-Brock and C. Grigson. Oxford: British Archaeological Reports International series 163.

A respectful and constructive reply by Glynn to Lewis Binford's assault on his and Mary Leakey's and their teams'work.

1984 The archaeology of human origins: studies of the Lower Pleistocene in east Africa, 1971-1981. *Advances in World Archaeology* 3:1-87.

A masterly overview of the indications of emergent human behavior discovered and presented in the ten years following Mary Leakey's great Olduvai monograph.

1985 Ancestors for us all: towards broadening international participation in paleoanthropological research. *Ancestors: The Hard Evidence*: 346-51. Edited by E. Delson. New York: Alan R. Liss.

Glynn's from-the-heart call for a de-colonization of research and archiving in his field.

1986 Foundation stones: early artefacts as indicators of activities and abilities. *Stone Age Prehistory: Studies in Memory of Charles McBurney*: 221-241. Edited by C.N. Bailey and P. Callow. Cambridge: Cambridge University Press.

Published posthumously in a tribute volume for one of his late great teachers, this is a succinct review of the searching out of ways stone tools may yield an account of emergent proto-human (but not yet human) patterns of subsistence.

in prep. *Koobi Fora.* Vol 5. Edited by B. Isaac. Cambridge: Cambridge University Press.